SPACEISM

THIS BOOK IS ABOUT RELIGION AND SCIENCE TO
DETER LIES THAT LEADS TO SINS AND CRIMES.

CZAR R. VALEN IS A FORMER TABLOID
REPORTER/PHOTOGRAPHER.

SPACEISM

A THEORY INSPIRED BY GOD/SPACE

BY

CZAR R. VALEN

EDITED BY:

ESTEBAN T. ROSALES JR.
RICTORINO G. ROCES JR.

"Your story is our priority"

LitPrime Solutions
21250 Hawthorne Blvd
Suite 500, Torrance, CA 90503
www.litprime.com
Phone: 1-800-981-9893

Published by LitPrime Solutions   05/23/2023

ISBN: 979-8-88703-243-6(sc)
ISBN: 979-8-88703-244-3(hc)
ISBN: 979-8-88703-245-0(e)

Library of Congress Control Number: 2023908480

CONTENTS

DEDICATION

I DEDICATE
THIS TO YOU

ACKNOWLEDGEMENT

Thank you for your precious time in considering this book. I am truly honored and humbled to share my thoughts hoping to excite your imagination. I also want to thank the following special people whom I have shared my thoughts and gave their comments while this book was still in conception.

"You might have a hard time with publishers"
(Dominador R. Estiva)

"You are just justifying what has become of you"
(Roberto D. Fajatin)

"You will confirm others and might convince some"
(The late Manuel T. Enerva)

"You mean there is no God?
(Luisito B. Pereira)

"Your title suggest that you have
no proof of your claims"
(Pantaleon V. Lapidario)

"Your Theory of Spaceism seems to make
sense but I must read the book"
(Carmelita A. Salandanan)

Thanks also to the New American Bible for the religious verses. The Webster's New Dictionary and online Dictionary.Com for the word meanings. Google/ Wikipedia for scientific research, Juliet A. Enerva for allowing me to use "THOUGHTS" written by her late husband, my childhood friend Manuel Tolaram Enerva and "DESIDERATA", Latin word meaning, "Things Desired" of unknown origin, but later known to be written by Max Ehrmann.

"Experience is the best teacher; books
are compilation of experiences, and the
Bible is a compilation of books."

This statement is true and reminds us that not all what is written are good and true. It also teaches us that knowledge could be acquired avoiding unnecessary loss of effort, wealth, time and worse, the loss of life.

ILLUSTRATION

Illustrations shall be provided by the reader while reading through its own imagination.

INTRODUCTION

I think humanity is founded on LIES that breeds all kinds of Sins and Crimes. Lie is not limited to the 9th Commandment of God, *EXSODUS: 20;16 You shall not bear false witness against your neighbor.* Lie is happening every time and everywhere that cause division among people and eventually becomes a humanitarian problem where all wants to be dominant blaming racism. To put an end to this, or at least reduced it to the lowest possible level, humans must unite under one unbiased God that will represent the whole mankind.

Some say there is only one God with different names in different religions. Some ignore the facts in science because of faith in religion. If this problem is not attended to as soon as possible, any invader, local or foreign could eventually succeed in replacing humans on earth and beyond.

Like this Corona Virus (COVD-19) pandemic outbreak that is killing people by the tens of thousands everyday all over the world. An enemy that knows no god, race, sex and age of a person. This virus is united in their objectives, that is to multiply, spread and kill. It is suspected to have been artificially created by evil people in a laboratory and spread like wildfire. Sadly, it seems that the fight is people against people and the only way humans could effectively fight back is for the people to unite.

Meantime, while waiting for any breakthrough in science for cure and vaccine against this deadly virus, prayers and simple guidelines were introduced to at least slow down the rate of infections and minimize the number of deaths. Discipline with proper hygiene, frequent hand washing and disinfecting, social distancing of at least six feet apart, wearing mask or face covering, avoiding indoor gatherings, testing, contact tracing and quarantine. But things got even worse when these simple guidelines were not observed by those who misunderstood religion and ignored science that only shows how divided humans are.

Religion and Science are supposed to be the source of all knowledge and contentment of mankind that depends on Ancestral Origin, Geographical Location and Environmental Influence on earth and beyond. Superiority or Inferiority only exist when one invades the space of the other because of Vanity and Greed.

I think, humans must understand and respect the Space between someone, something, anything, everything because nothing is possible without the Space.

According to science and history, the Black People of Africa were the first to roam the surface of the earth. Is God black? In Chromatics, the study of colors, black is not a color but the absence of color and white is the combination of all colors. I heard a story how the people around the world acquired different skin colors and the story goes like this.

During the creation, God used a special kind of pure white clay to mold and shape humans according to God's own image and likeness. God made several sets of man and woman using the white clay as mold, confidently baked them without using the timer. They got burned and turned black, some got reddish in color and God placed them where they belong. On the second sets of man and woman, God used the timer, but was set too early that it was not cooked well. Some were still white, and some got yellowish in color, then God placed them where they belong. However, on the third sets of man and woman, God did not use the timer but watched and checked them every now and then until the desired color was reached. Not black, not white, but brown, the perfect color of humans and God was pleased, placed them where they belong, then life begun.

Of course, the story is obviously a mere conception of the human mind just like Superman, King Kong, Tarzan, Wonder Woman, Spiderman, Captain America, Hulk, King Arthur, Harry Potter, Planet of the Apes, Star Wars, Star Trek and many others that shows how amazing the human mind works.

Like the story of Naga City in the Philippines, the place where I was born. I think, when the history of Naga City was written, the word Naga was localized and made it beautiful compromising some truth.

According to the writer, Naga derived from the word Narra, a kind of tree that was abundant in the place during the early days. Surprisingly, there is not a single Narra tree in the place today. Maybe when Naga was urbanized, they cut down all the trees up to the last Narra tree standing.

The word Naga is a Snake Goddess (Sanskrit Serpent) half human half serpent. I think, during those days, the place was infested with snakes that the residents with Hindu origin named the place Naga in honor of the Snake Goddess. Evidence of this is the famous Philippine Cobra Snake that are found in the adjacent town Haring. The word Haring could have meant to run fast.

Perhaps, when Naga was booming, people are coming to settle down in the place for trade and commerce. The snakes hurriedly migrated to the nearby place later named Haring. The word Naga was deeply rooted in the minds of the residents that when the story of Naga was written, the word Naga was maintained, but the meaning was localized and matched with a beautiful fictitious story.

I think all on earth and beyond, be it material, event, or concept is a cycle, a chain of related events that end and begin at a certain point in time. I am focusing on the beginnings not because first impression is lasting but to make sure that truth is told in the beginning to ensure a better beginning after the end. As the saying goes, "History Repeats Itself", this way, events will have a chance to be perfect. As for life, it will undergo series of rebirth called resurrection and reincarnation until the soul is purified and ready to enter heaven in union with God in the everlasting life and have perfection again.

I was born, baptized, and raised as a Roman Catholic and fascinated in the life story of Jesus the Nazarene called the Messiah. I do not understand why there are those who change their religion when they can choose between good and bad even without religion as what Adam and Eve did in the garden of Eden. Some say because of salvation that will ensure the soul of a departed to be with God in the afterlife. But what if God and the afterlife are mere conception of the human mind?

As a boy growing up, some old folks in our neighborhood told me to handle the Bible with utmost care by placing it on top of anything else because this is where the words of God were written. Out of fear without understanding why, I did what I was told. Even prayer books and other things that has something to do with religion and God, I treat them the same way I treated the Bible.

I was taught that masturbation is a sin and even by just thinking of sinful act, one has sinned already. I was also taught that anything new must be blessed first in church during the Holy Mass before using them regularly. When the church bells tolls at 6:00 in the afternoon for Angelus, we must make the sign of the cross. I also remember, we used to pray the Rosary in front of an altar in our house that has a three-foot-tall cross with the image of Jesus the Christ nailed to it. There were also other images and religious reading materials for different occasions. As much as possible we used to eat together and pray to give thanks before every meal.

My understanding in religion and God gradually changed when I started questioning my belief. I have nothing much to be proud of my childhood specially when I become curious of my surroundings and started doing things out of the ordinary. Things got even worse when I learned the art of lying, doing wrong and bad things getting away with it by lying. Eventually I got into trouble and was transferred from a Private Catholic School to a Public Non-Catholic School and was branded as the black sheep of the family. This made me an introvert, a loner, often gazing into the sky watching the stars while laying down on the roof of our house when the sky is clear during the early time of the night and often wonder where God is. I think, being with God in the afterlife is the greatest thing that could ever happen to a human soul.

In my early twenties, as I lay on my bed resting, I saw red writings on the wall.

WITH SINCERITY YOU ASK FROM THEE AND SO IT SHALL BE, NATURE IS BUT THE REFLECTION OF THYSELF.

I closed my eyes and when I look again, the writings on the wall were gone. Maybe I was so exhausted that day due to the long religious discussions we had during the night that made me fall half asleep and had a dream. Or maybe it was due to the doubts accumulated in my mind that made me questioned my faith in God.

I sat at the edge of the bed, closed my eyes, and saw the writings in my mind exactly how it was on the wall. I tried to figure out what it meant but could not satisfy myself. Since then, I decided to share it to friends and relatives, hoping to find the meaning that would satisfy me. But instead, I often find myself ridiculed and asked the question, "Why you"?

Sometime in the 1980's, my childhood friend Manny, the late Manuel Tolaram Enerva, introduced me to a group called, the Theosophical Society in the Philippines (TSP). Although I did not join the group that time, Manny and I often see each other and talked about religion. Before Manny passed away in February 2014, I joined TSP sometime in 2013 and started attending seminars. One day, I went to the TSP Headquarters and saw a poster that says, "THERE IS NO RELIGION HIGHER THAN TRUTH". Later I found out that it is the slogan of the group.

I was challenge by this and decided to search for this "TRUTH". I started sharing it to friends and relatives, but the question keeps coming back to me. What is that "Truth"? Out of frustrations I found myself drawn towards the mystic serenity of the Space that made me compare God and Space that eventually led me to my "Theory of SPACEISM".

There was this girl friend of a friend who asked me, "Do you know your purpose in life why you live?" I said "yes" and replied, "I live to die". After that incident, she never talked to me ever again. Honestly, as I am writing this now, I still do not know what my purpose in life is. I hope you will not walk out on me just like what she did. Please be patient while I share my thoughts without boring you with details by KISS (Keeping It Simply Short).

NATURE

Nature is the essential characteristics and fundamental qualities of all that are Visible and Invisible on earth and beyond with their forces, inhabitants and all its phenomena not made or created by humans.

Among the estimated two trillion galaxies in our universe, our galaxy with trillions of planets, only planet earth is known to be inhabited with living things. Science categorized living things into two kingdoms, the Animal Kingdom and the Plant Kingdom. But modern science categorized it into five kingdoms: Animal, Plant, Fungi, Protist and Monera. According to science, Mankind is the highest form of living thing in the Animal kingdom and in all the Kingdoms because only Man has a brain that is capable of highly complex mental activities.

All major religions believe that living things have two bodies, the Visible Physical Body and the Invisible Spiritual Body. An individual person has an organ called brain that is protected by the skull of the head because it is the central motor that is responsible for all the functions and activities of an individual person. A healthy and sound brain is a healthy and sound person.

I think, the brain does not think but respond

to all the five senses of a person, Sight, Taste, Smell, Hearing and Touch by instinct. Instinct is the inborn natural impulse of the brain to all the senses. The brain functions like a hub that receives, stores, process and relays chemical and electrical signals to all the internal and external parts of the physical and spiritual bodies of a person.

Mind is the spirit element in the brain of a person that do the thinking, brain of the soul. It is the total consciousness that has the freewill to think and responsible for all the actions and decisions done by a person.

Memory is the total stored data in the brain that can be recall anytime if needed. Data is the information's accumulated stored from the functions and activities of a person that do not die. I think, Memory is the other self or the body of the Soul that can be judged and remembered after a person dies.

In the sixties, Dr. Paul MacLean an America physician and neuroscientist introduced the concept of Triune Brain. According to him, the Brain structure and function is based on three specific regions of a person, the Head, the Heart and the Gut. I think, when the mind thinks and the brain send signals, the Heart and the Gut are affected as recipient. The Heart could be affected by emotions and the Gut could be affected by hunches, but both Heart and Gut has no brain that stores data in memory and has no mind that think of its own.

Christians believe that when a person ceases to live or dies, the Physical Body returns to nature, "ashes to ashes, dust to dust"; and the Spiritual Body returns to God, the breath of life but the Soul lives on to be judge in the afterlife.

> *GENESIS: 2;7 the Lord God formed man out of the clay of the ground and blew into his nostrils the breath of life and the man became a living being.*

> *ECCLESIASTES: Chapter 12;7 And the dust returns to the earth as it once was, and the life breath returns to God who gave it.*

Knowing that death was inevitable, the Mind created the Soul, the Memory that will live in the Afterlife. The Memory or Soul lives after a person dies to be judged in the afterlife according to the quality of life the person has lived to determine where the Soul will go. If good, the Soul goes to heaven in union with God in the everlasting life. If bad, the Soul goes to hell in eternal damnation in the everlasting fire. The Memory of the departed one that will be remembered in the minds of the living.

Adam and Eve disobeyed God and acquired the knowledge of good and bad, but mankind lost the everlasting life.

> *Genesis: Chapter 3;22 Then the Lord said: "See! The man has become like one of us, knowing what is good and what is bad! Therefore, he must not be allowed to put out his hand to take fruit from*

the tree of life also, and thus eat of it and live forever."

Genesis: Chapter 3;7. Then the eyes of both were opened, and they realized that they were naked; so, they sewed fig leaves together and made loincloths for themselves. 3;8. When they heard the sound of the Lord God moving about in the garden at the breezy time of the day, the man and his wife hid themselves from the Lord God among the trees of the garden. 3;9.the Lord God then called to the man and asked him, "Where are you?" 3;10. He answered, "I heard you in the garden, but I was afraid, because I was naked, so I hid myself." 3;11. Then he asked. "Who told you that you are naked? You have eaten, then, from the tree of which I had forbidden you to eat!"

From then on, mankind lived in fear, but God so loved his creations that when he drove Adam and Eve out of Paradise, God see to it that communication between God and mankind shall continue. God gave them a built-in wireless transceiver called Conscience for communication. It receives all data but transmit only positive or good signals what we call "the God within". Conscience guides the Mind of a person to think good and do the right action. When the mind of a person goes against the Conscience, that person will feel guilty, but the Conscience will never go against a person.

Conscience is pure and an unadulterated inner essence of a person that never lies. It is a combination of two words, Con and Science where Con means against or in defiance and Science is the systematic method of proving facts. Therefore, Conscience separates good from bad and protects the person by guiding the Mind to do the right action. Without the Conscience a person cannot be humane as a human being and will only be a plain and ordinary animal that will grow and develop by instinct.

All thoughts, imaginations and ideas of the Mind are spirits that can manifest in a state of things not actually present but comes into being through all forms of medium. They are transformed into real tangible visible material things such as buildings, roads, bridges, vehicles, ships, airplanes, foods, shoes, cloths, perfumes, jewelries and many others. Others are transformed into tangible visible material things but are not what they are. Like Demons, Angels, Ghosts, Fairies, Wonder Woman, Superman, King Kong, Chucky, Spiderman, The Hulk and the likes.

I think, Demons are negative spirits that manifest out of evil, fear and anxiety. Angels are positive spirits that manifest out of virtue, beauty and kindness. Ghosts are spirits that manifest as appearance or semblance imagination of things that have ceased to exist. All that is associated with the Ghost can manifest in all the five human senses through imagination and emotions. All these manifestations will linger in the minds as knowledge stored in the Memory in the brain that will be handed down from generations to generations to be re-tolled again and again and again. A Legacy or the Life-After-Death of a person that will be remembered over and over and over again.

I think, an unborn baby is already equipped with the Physical and Spiritual Bodies while still in the womb during pregnancy. The Material Physical Body starts learning and growing while still in the womb through the brain and the five senses by instinct. I think, in this stage the baby is not yet a complete human being. When the baby is born and take the "Breath of Life" for the first time, the baby becomes a whole living person. When that person starts to think and reasons, Conscience starts to work, and the person becomes a human being. Then the Soul (memory) starts to grow through the thinking of the Mind.

I think, all living things have a brain of its own that respond through all its senses by instinct but do not have Conscience. Only humans have Conscience that when a person has done something wrong or bad, the Mind will be bothered by the Conscience.

When a person is in danger or about to do something bad or wrong, Conscience turns on to question the action to be taken. It will automatically pullout files from the memory in the Brain to be decipher by the Conscience. After the Conscience worked on it and relayed the message to the Mind, the Subconscious-Mind decides, then dictates the Conscious-Mind to act.

When the Subconscious-Mind do not agree with the Conscience, the Subconscious-Mind overrules the Conscience to justify the action, then dictates the Conscious-Mind and the person lies. If the Subconscious-Mind agrees with the Conscience, the Subconscious-Mind dictates the Conscious-Mind, and the person do not lie. It is the Conscious-Mind in collaboration with the Subconscious-Mind of a person that lies, but never the Conscience. If we go against our Conscience, we must be prepared to answer for whatever corresponding consequences there might be.

Conscience can automatically decipher anything that pops out in the Human-Mind. When the result is divine, it is called Intuition, when it is evil, it is called bad Omen. If Conscience discerns Intuition further with reasons and reveals its hidden truth, it is called Insight.

Intuition is the immediate and direct apprehension of the Human-Mind without reasoning. This is deceiving and dangerous because it is not clear where it is coming from. On the other hand, Insight is the mental penetration of reasons from our Conscience to gain accurate and deep intuitive understanding in search for truth. I think, Insight is the ability of the Human-Mind to revel hidden truth in Intuitions, "soul searching". It is better to think twice than to be impulsive and be sorry.

Freewill is the voluntary freedom of choice of a person between good and bad. Although sometimes there are instances that things would seem to be destined, we often say, "it is God's will". I think, it is not God's will because this so-called predestined happening happens when earlier factors that one had chosen lead to such. There is always the choice between good and bad, right and wrong but sometimes we chose contrary to what our Conscience dictates. When we go against our Conscience, it is a bad decision. Whatever happens, it is our final decision or choice that counts. Therefore, for best results, it is always wise to remember Karma and consult one's inner self, the Conscience.

I think, everything on earth and beyond, be it material, concept or event are all symbiotically in cycle of unending cause-and-effect. Waters on earth evaporates, condensed, and fall as rain to complete the cycle. Carbon dioxide as human waste is absorbed by plant's leaves as food then emits oxygen as plant's waste for humans to breath.

The lands and waters could drastically change in every major earth's cycle, where one civilization ends, and another begins. When the sun shines on earth, with all the essential ingredients for life, life on earth continues with a different landscape but life on earth will still be the same.

Although earth and everything on it and beyond is material, we cannot do away with the spiritual. I think, what humans must be concerned of now, is how to protect, improve and maintain the ecological balance of the earth and beyond. I think, earth is our universal mother, and the sun is our universal father that makes all humans universally brothers and sisters. Our so-called biological parents are merely instruments for the process called birth. Thus, NATURE (earth) IS BUT THE REFLECTION OF THYSELF (human).

If Matter is a substance made up of various types of particles that occupies physical space and has inertia; and Space is a boundless, three-dimensional extent in which objects and events occur and have relative position and direction. I think, Matter cannot exist without Space, even Space cannot exist without the Space. God created all that is in the Space, but did God create the Space?

WORD

Word is a single symbol composed of one or more spoken sounds used in speaking and writing in a language as a system of communication. Perhaps word existed since the beginning of life in a simple sound with a simple gesture, like a newborn baby and evolved into what we called word.

The consensus is that the Sumerian was the first written language developed in southern Mesopotamia around 3400 or 3500 BCE (Before Common Era) or BC (Before Christ). Sumer was an ancient civilization founded in the Mesopotamia region now Iraq, situated between the Tigris and Euphrates rivers. I think, it is important to note what the Bible say about the Word.

> *John: Chapter 1;1. In the beginning was the Word, and the Word was with God, and the Word was God. 1;2 He was in the beginning with God. 1;3 All things came to be through him, and without him nothing came to be. What came to be 1;4 through him was life, and this life was the light of humans; 1;14 And the*

word became flesh and made his dwelling
among us, and we saw his glory, the glory
as the father's only son full of grace and
truth.

These verses are not in the Old Testament of the Bible, they are in the New Testament. Although the Book of John the Apostle is not considered a synoptic like the Gospel of Matthew, Mark and Luke that are closely similar in contents and wordings, it is accepted as a gospel because it talks about the life and death of Jesus the Nazarene.

I think, probably, when John the Apostle was writing his book, inspired by God, he saw the need for God to use Word where Jesus fits in. Otherwise, God would have been mute, and creation might have not taken place. The first word God uttered in the Bible was, "Let", meaning, to allow.

> *GENESIS: Chapter 1;3 Then God said,*
> *"Let" there be light,*

Science claims that several tests showed that words can influence plants. One experiment done was placing in separate rooms the same variety of plants, one spoken with kind words and the other with harsh words. The results shows that the plants spoken with kind words grew a little more than the other spoken with harsh words. This shows that word influence other living things.

Humans have the power to create or destroy anything or anyone using a word or words. It can send a person to prison or worse to death even if not guilty of any wrongdoing and can set free a person who is guilty of wrongdoings. Since then and until now, whenever there is a new thing, concept or event that is without a name, humans have the means and capability to give a name for such. It is important to remember not to twist or misuse words because it will give confusing meaning instead of understanding. This twisting and misusing of words are deliberately done by people with bad intentions (liars).

I think, the word LIE is the root of all evil. Wealth, Fame, and Power are good status symbol of a person but can become evil when it is a product of Lie. Lie in the Bible can be traced back, when God favored Abel's offering that made Cain killed Abel and tried to get away with it by lying.

Genesis: 4;8. Cain said to his brother Abel. "Let us go out in the field". When they were in the field, Cain attacked his brother Abel and killed him 4;9 Then the Lord asked Cain, "Where is your brother Abel?" He answered, "I do not know. Am I my brother's keeper?"

When this incident happened, the Ten Commandments of God was not yet known to mankind but two of these commandments occurred in this event, Killing and Lying.

> *EXODUS 20;13 You shall not kill. 20;16 You shall not bear false witness against your neighbor.*

A person who lies can do all kinds of wrongdoings

including all the rest of the Ten Commandments of God and could get away with it by lying. Lying is done with bad intension to misled someone by false statement that can have serious consequences. If lying is done frequently, it becomes a habit and can become a disease, Pathological Lying. Pathological Lying is the chronic compulsive behavior or habitual lying about anything big or small in any situation and even lies to oneself. Lies with good intentions like saving someone from guilt are called white lies. But lies are lies, whatever, they are still lies and guilt is still guilt. Remember, "Big things comes from small things", "One cannot have a dollar without a penny".

I think, lying must be criminalized and liars must be given the stiffest punishments. Perhaps, Lawmakers could create laws that will deter people from lying because it is so rampant that almost everyone, everywhere, and every time lies that cause confusions, destructions and deaths. We must stop lying and start using words honestly and respect the words of others. Say what you mean and mean what you say, "Word of Honor". A liar is the worse and most dangerous person on earth. As the saying goes, "Honesty is the Best Policy."

Bible Authorities assumed that God have spoken to Adam using the Adamic language introduced by Adam in the Garden of Eden.

> *Genesis: Chapter 2;19. So the Lord God formed out of the ground various wild animals and various birds of the air, and he brought them to the man to see what*

*he would call them; whatever the man
called each of them would be its name.*

When we pray, we use words in our thoughts and
when we pray to God done earnestly and repeatedly
in deep concentration, it becomes self-hypnosis. I
think, under hypnosis the brain emits positive and
negative electromagnetic forces that travels through
electromagnetic waves. Electromagnetic force is the
fundamental force responsible for atomic structure
and chemical reactions that charges all phenomena.
When electromagnetic force combines with other
electromagnetic force during prayers, it produces results
(answered prayer). This result can be positive or negative
depending on the intention, situation and how it was
delivered to produce the kind of result expected. For
example, when a person prays for the healing of another
person's sickness and the sickness gets well, the prayer
was answered positively. But if the sickness did not go
away, the prayer was answered negatively or was not
answered at all.

"It is better to give than to receive". This statement is
unfair because it is not reciprocal "give and take". When
giving without expecting anything in return, it is help.
Otherwise, it is better to receive because the subject is
given not taken. Taking is wrong because it is not given
freely especially if it is done with force.

"First impression is lasting". This statement is harsh
because it has no room for change and improvement.
Whatever the person's first impression will stick and last.

"Don't judge a book by its cover". This statement

gives a book a chance to be read or a person a change to improve even not having an impressive first impression.

"The end does not justify the means". Remember, answer always comes after the question. If answer is the end and question, is the means. Therefore, the end thus justifies the means.

"Avoid loud and aggressive persons; they are vexation to the spirit". Compare the two person and tell who vexation to the spirit is. John the Baptist, with loud and aggressive voice telling the people to repent. Or a person who is whispering to his companions his plan to kill another person.

It is said that actions speaks louder than words, but words speaks clearer than actions. I think, to lessen the problem of miscommunications and misunderstandings among humans, "Ask and Accept", a policy that must be practice and observed. "Ask" freely whatever question is in the mind and "Accept" with respect whatever the reply.

Medieval Jewish maintained that the Hebrew language was the language of God during creation until the destruction of the Tower of Babel. During this time Nimrod was king and mankind was united speaking in one tongue but what happened?

> *Genesis: Chapter 11;5. The Lord came down to see the city and the tower that the men had built. 11;6 Then the Lord said, "If now, while they are one people, all speaking the same language, they have started to do this, nothing will later stop*

them from doing whatever they presume to do. 11;7 Let us then go down and there confuses their language, so that one will not understand what another says." 11;8 Thus, the Lord scattered them from there all over the earth, and they stopped building the city.

Clearly this God does not want to be found and be understood. I think, maybe this Tower of Babel is a metaphor for the highly advance technology they were able to achieve during their time. Perhaps, King Nimrod became greedy and ambitious that he wanted to be God himself that lead to the catastrophic end of their highly advance civilization and resulted to language confusion once again among the people. Perhaps, this event was part of an Earth Cycle. "History repeats Itself" that gives mankind the chance to make things right in the next cycle to come. I think, peace on earth will only be possible if humanity will once again unite speaking in one tongue without Vanity and Greed under one unbiased God.

RELIGION

Religion is a system of faith and worship to a Supernatural Being or God.

Faith, because God is an eternal spirit entity that exists in the minds of the believers. An invisible God that has no beginning and has no end who created everything.

> *HEBROS: Chapter 11;1 Faith is the realization of what is hope for and evidence of things not seen.*

> *ISAIAH: Chapter 40;28 Do you not know, or have you not heard? The Lord is the eternal God creator of the ends of the earth. He does not faint nor grow weary.*

Some estimates that since the beginning of time up to this date, there are roughly Four Thousand Two Hundred (4,200) religions all over the world and still counting. Religion is important to humanity because they are supposed to be the source of spiritual morality, contentment and peace of mind of the believers. There are those who believe that there is only one God with different names in different religions.

Christianity and Islam are the two largest religions in the world today. They both originated from Judaism that started in the Middle East and shared historical connections, with some major theological differences. Judaism is an ethnic religion of the Jewish people, the covenant that God established with the children of Israel.

According to scholars, God first revealed himself to a Hebrew man named Abraham, who became known as the founder of Judaism around 9th – 5th century BC. Judaism, Christianity and Islam all recognized Abraham as their first prophet and they are also called the Abrahamic religion.

Christianity originated in the Roman province of Judea founded on the life and teaching of Jesus of Nazareth. Jesus is also called Christ, meaning "Messiah" or "Anointed One". Jesus is the son of Joseph the carpenter and Mary the Virgin and those who follow him are called Christians during the 1st century AD (Anno Domini) Latin phrase meaning, Year of the Lord or CE (Common Era).

There are skeptics that do not believe that Jesus did exist, but his life was the greatest story ever told and his greatness is deeply embedded in the minds and hearts of the believers. Whether Jesus did exist or not, he is the source of knowledge and wisdom.

Islam started in Mecca, modern-day Saudi Arabia, during the time of the prophet Muhammad. Those who followed it are called Muslims which means "submitter to God" that developed during the 7th century AD.

Since then, Christians and Muslims conflicts have destroyed countless properties and lives that is still happening until now. Like the ten-year-old civil war in Syria that is believed to be caused by poverty and mistreatment of the Syrian government to its people. I think, it has something to do with religious rivalry in the country and in other nations in the Middle East. Another is the Israel and Palestine conflict that could also be caused by religious differences.

All known and unknown, visible and invisible, virtual and actual, physical and spiritual on earth and beyond revolves around religion in fear as a system of faith and worship. In the Bible, when God drove Adam and Eve out of the Garden of Eden, they realized their mistakes and the need for forgiveness that made them start worshiping God.

> *Genesis: Chapter 3;8. When they heard the sound of the Lord God moving about in the garden at the breezy time of the day, the man and his wife hid themselves from the Lord God among the trees of the garden. 3;9 The Lord God then called to the man and asked him, where are you? 3;10 He answered, I heard you in the garden; but I was afraid, because I was naked, so I hid myself.*

Fear, because God is a Jealous and Vengeful God, who punish those who do not obey His will where the innocents are not spared. This is worse than rival religions killing each other to prove that their God is more superior and powerful than others.

Jealous.

> *EXODUS: Chapter 20;4 You shall not carve idols for yourselves in the shape of anything in the sky above or on the earth below or in the waters beneath the earth; 20;5 you shall not bow down before them or worship them. For I, the Lord your God, am a jealous God, inflicting punishment for their father's wickedness on the children of those who hate me, down to the third and fourth generation.*

Vengeful;

> *ROMANS: Chapter 12;19 Beloved, do not look for revenge but leave room for the wrath; for it is written, "Vengeance is mine, I will repay, says the Lord."*

> *GENESIS: Chapter 6;7 So the Lord said: "I will wipe out from the earth the men whom I have created, and not only the men, but also the beasts and the creeping things and the birds of the air, for I am sorry that I made them."*

Is this the just, merciful and loving God? A God with corruptible human traits that could lead to chaos, destructions, and deaths? I think, this is the result when God humanized himself and made man divine during the creation of man.

> *GENESIS: Chapter 1;26 Then God said: "let us make man in our image, after our likeness.*

The Holy Bible is said to be a "Book of Books" inspired by God and written by men.

> *THE SECOND BOOK OF SAMUEL: 23;2 The spirit of the Lord spoke through me; his word was on my tongue.*

Saint Paul the Apostle, who spread the teachings of Jesus is not one of the original Twelve Disciples but considered the most important person in the history of Christianity after Jesus during the First Century. He founded Christian communities and send letters to different congregations that had enormous influence on Christian theology. He met Saint Timothy during one of his travels and wrote him this.

> *2 TIMOTHY: 3;16 All Scriptures is inspired of God and is useful for teaching for reproof, correction, and training in holiness 3;17 so that the man of God may be fully competent and equipped for every good work.*

Saint Timothy was born of a Jewish mother and

become a Christian believer and that people spoke highly of him. Saint Paul and Saint Timothy were both imprisoned because of their faith.

Although Christianity and Islam are the two largest religions in the world today that both originated from Judaism, they are not the first religion that existed on earth in recorded history. According to most of the scholars, the oldest religions that existed on earth in chronological order are as follows.

1. Hinduism - is the world's oldest religion in recorded history that originated from the Indo-Aryan people that migrated to the Indus Valley and blended their culture with the indigenous people living in the region around 15th – 5th century BC (Before Christ) or BCE (Before Common Era).

2. Zoroastrianism - is the second in recorded history and known to be the official religion of ancient Persia now Iran founded by the prophet Zoroaster in 10th – 5th century BC.

3. Judaism – is the third in recorded history, an ethnic religion of the Jewish people, the covenant that God established with the children of Israel and where Christianity and Islam originated from around 9th – 5th century BC.

4. Jainism – is somewhat like Buddhism a rival in India. It was founded by Vardhamana Jnatiputra or Nataputta Mahavira called

Jina (Spiritual Conqueror) contemporary of Buddha around 8th - 2nc century BC.

5. Confucianism – is a system of ethics devised by the Chinese scholar K'ung Fu-tzu (Latinized to Confucius, a great teacher but not worship as a personal god. Nor did Confucius himself ever claim divinity, founded around 6th - 5th century BC.

6. Buddhism – arose in the eastern part of Ancient India based on the teaching of Siddhartha Gautama now known as "Buddha" around 6th - 5th century BC.

7. Taoism – is a Chinese philosophy attributed to Lao Tzu and became the official religion of the country under the Tang Dynasty around 6th – 4th century BC.

Hinduism is an Indian religion or way of life that is now the third largest religion in the world. I think, Judaism, Islam and Christianity was easily embraced by the people when it was first introduced because it is less complicated. It adheres to only one God unlike Hinduism that worships many Gods and Goddesses with Brahman as the supreme God responsible for the creation of the world and all the living things.

Hinduism believes in Karma meaning action, that good action gives good result and bad actions gives bad result in the cycle of life, death, and rebirth. The concept of Karma began with the question about how and why man is born and what happens after death. Karma first appeared in the oldest Hindu text the Rigveda before the 15th BCE. It is a Sanskrit word that primarily means action-and-result known as a Natural Universal Law of Karma. It is also referred to as the spiritual principle of cause-and-effect which states that an action is always accompanied by its consequence or equal reaction. I think since the beginning of time until now all on earth and beyond are series of cycles of an unending cause-and-effect that good or bad actions determines the future of an individual person that is similar to the Christian's, "we reap what we sow".

> GALATIANS: 6;8 because the one who sows for his flesh will reap corruption from the flesh, but the one who sows for the spirit will reap eternal life from the spirit.

Young Earth creationism (YEC) believed that Earth and its lifeforms were created in their present forms thru God's Grace. A supernatural act of the God of Abraham approximately between Six Thousand (6,000) and Ten Thousand (10,000) Years Ago.

Some scholars say that Paganism is as old as mankind that started with Cain, the son of Adam and Eve who refused to worship God. If this is true, Paganism known to be the "religion of peasantry" must be older than Hinduism and could be the oldest religion in recorded history on earth.

> Genesis: Chapter 4;3 In the course of time Cain brought an offering to the Lord from the fruit of the soil, 4;4 while Abel, for his part, brought one of the best firstlings of

his flock. The Lord looked with favor on Abel and his offering, 4;5 but on Cain and his offering he did not. Cain greatly resented this and was crestfallen. 4;6 So the Lord said to Cain: "Why are you so resentful and crestfallen? 4;7 If you do well, you can hold up your head; but if not, sin is a demon lurking at the door: his urge is toward you, yet you can be his master."

Paganism is a broad term with various meanings, it is noted that in some contexts it was used to mean any non-Abrahamic religion. Often Pagan was used to mean something along the line of "infidel, heathen, depraved, false, wrong, lying and even devil worship".

Abraham, the forefather of Jesus Christ was a Pagan. Did God bribe Abraham that corrupted his mind to make him renounced his Pagan way of life and become faithful to God?

Joshua: Chapter 24;2 Joshua addressed all the people: "Thus says the Lord, the God of Israel: In times past your fathers, down to Terah, father of Abraham and Nahor, dwelt beyond the river and served other gods. 24;3 But I brought your father Abraham from the region beyond the river and led him through the entire land of Canaan. I made his descendants numerous and gave him Isaac.

I saw a boy about twelve-years-old in a food distribution center helping load boxes in a vehicle. Then a lady, probably the owner of the vehicle, handed the boy two dollars in paper bills. The boy was surprised and did not know what to do but the lady insisted. The boy took the money and said, "thank you". Surely the boy learned something that day. I think, maybe in the beginning, bribery and corruption starts with no bad intentions. Once the bad intention is present, it becomes a Sin and a Crime.

I think, Paganism is a combination of religion and government. In the Bible, around 306 to 337 ADS, the Pegan emperor Constantine the Great, (Flavius Valerius Constantinus) issued an order that protects the Christians from persecution. Emperor Constantine the Great was the first Emperor of Rome to convert into Christianity. Under his rule, Roman religion (Paganism) and Christian religion (Christianity) merge into one and the Roman Catholic Religion was born. It can be traced to the life and teaching of Jesus Christ in the Roman occupied Jewish Palestine about 30 AD. According to the Roman Catholic teaching, each of the seven Holy Sacraments Baptism, Eucharist, Confirmation, Reconciliation, Anointing of the Sick, Matrimony and Holy Orders were instituted by Jesus himself. The purpose of the sacraments is to sanctify men, to build up the body of Jesus, and finally, to give worship to God (Second Vatican Council).

Later, church and state were separated because of vanity and greed. Religion was under the church and the government was under the state. Later both were corrupted and deflected with greed for wealth and power. Many religions and governments were formed to have a piece of the pie. They too were corrupted and were used with bad intentions that gets us into what and where we are now. All religions and governments will start united with good intensions but end up corrupt that breeds new ones only to end up the same. Vanity and Greed breeds Divisiveness among religions and dominate one another claiming to be superior and the true way to salvation.

Religion is the unifying factor and the source of spiritual morality, peace, and contentment of the believers. Therefore, the government must seek religious guidance to find the TRUTH without Lies. When all the peoples of the world will unite once again, speaking in one tongue just like the time of Nimrod, with all its corrections and changes. The new civilization will begin with better understanding of life and will worship God without fear. Then earth will once again be a better place to live in, as it was in the beginning, now and ever shall be world without end, amen.

SCIENCE

Science is the systematic method of acquiring true facts through observations and experimentations. It is a fundamental structure in education and an integral part for maintaining the Ecological Balance of the earth and beyond.

Technology is the branch of science that deals with the invention and creation of gadgets and chemicals with the use of knowledge and the earth's natural resources in relation to life, society and environment. According to scientists' technology can damage the earth in two ways, pollution and depletion of its natural resources.

Depletion is the removal or reduction of earth's natural resources that aside from pollution, it could also erode the soil and move the rocks underneath the ground that can trigger Earthquakes and Tsunamis. Pollution are toxins from technology that contaminates the food that we eat, the waters that we drink, the air that we breath and nature itself.

Studies shows that aside from damage to infrastructures, air pollution alone cause more than three (3) million deaths each year around the world. Remember, "Nature is but the reflection of thyself", where earth and beyond is nature and what we do to nature is what we do to ourselves.

Most climate scientists agree that today's average global temperature is warmer than in the pre-industrial times because of Global Warming that is attributed to the various pollutants from technology.

Global warming is the increase in the earth's average atmospheric temperature that causes corresponding changes in climate that may result from Greenhouse Effect in maintaining the right temperature from the sun to sustain life. However, the increase in the absorption by atmospheric carbon dioxide is resulting to a concerning level.

The United States foremost agencies and organizations that are involved in climate science have recognized Global Warming as a human caused problem that should be addressed. There are groups that blames Population Explosion as the primary cause of Global Warming.

Population Explosion is the sudden increase in the number of people in a population. In a study, global increase in human population is estimated to be around Eighty-Three (83) Million yearly or One-Point-One Percent (1.1%); and the global population has grown from One (1) Billion in the year 1800 to Seven-point-Nine (7.9) Billion in the year 2020.

I think, earth will never be overpopulated because all living and non-living things existing with all the natural resources, no more no less, dead or alive, stays on earth and beyond. Human population before is not far from what it is today because people before are scattered that seemed to be fewer but now that they tend to get together seems to be increasing.

I think, all on earth and beyond is a cycle within a cycle. Population is a cycle that will start and end at a certain number of people. Maybe in some areas population is increasing fast but in some areas the population could be slowly increasing or maybe none. Birth and Death balances the population of all living things that the total population worldwide will remain in the same average because there is nothing added to or subtracted from the earth and beyond. If things will all be in its proper perspective, Global Warming and Climate Change will still exist as part of earth cycle, where life expectancy of all living things might change with Birth and Death still playing its role to sustain life on earth and beyond.

Science and Technology is supposed to help humanity and protect the earth and beyond through Faith in God and Newton's Universal Law of Causality "Cause and Effect", the cycle of Life, Death and Re-Birth. When Lies increases, sins and crimes increases that creates humanitarian problems. This makes humans to be racist and divisive. Then Vanity and Greed make humans dominate one another using technology, blaming religion and science.

George Lemaitre is a Belgian cosmologist and a Catholic priest who introduced the Big Bang theory. According to him the universe began with a single Primordial Atom that exploded and scattered what is now known to be the space and the whole universe. It is recorded that many and even scientists were forced to accept blindly some of the inconsistencies in this theory. Like the magnetic monopoles, the flatness, the matter and antimatter and even viewed as an alternative to God's creation in the Bible. If this theory is true, where was this Atom before the explosion? I think, the Atom did not create the space, but the Atom was in the space when it exploded creating the universe. Or perhaps the universe comes out, grow, sprang, originate, emanate from space.

In 2012 scientists claims that they have discovered the Higgs Boson also known as God Particle. The Higgs boson according to science is the fundamental force carrying particle of the Higgs field, which is responsible for granting other particles their mass, found at the Large Hadron Collider, the most powerful particle accelerator in the world located at the European particle physics laboratory CERN, Switzerland. This field was first proposed in the mid-sixties by Peter Higgs, for whom the particle is named and his colleagues.

Universe is the totality of supposed known objects its forces and phenomena in Space. Scientists agrees that the universe is about Thirteen Point Eight (13.8) Billion Years Old, and the earth is estimated to be Four Point Fifty-Four (4.54) Billion Years Old. According to science, Space is a boundless, Three-Dimensional extent in which objects and events occur and have relative position and direction. Truly the human mind is mysterious and fantastic but only God and Space knows when and how it all happened.

It is possible that life do exist in another planet like earth somewhere in the very far universe in Space.

Or it could also be a misconception of the human mind because until now it is still a possibility with no concreate evidence. I think, if ever there is a planet like earth out there in the far universe, Space must have positioned them in such a way that it will have no conflict with our planet earth. Unless one of the planet's inhabitants becomes greedy and ambitious to invade and dominate others and unless they have created refueling and recharging stations in the universe to sustain the travel.

Do we really need to search for another planet like earth when we already have one? If Adam and Eve did not disobey God's order, there will be no death. Earth and beyond could now be overflowing with living things because there is no Death to balance Birth. Earth's destruction or end of the world could only be a part of the earth cycle where people in previous civilization has predicted its near extinction that made them search for God and another planet like earth as an alternative destination for life to continue. If ever we do find one, I am sure it is occupied, and we will be intruding. Why not just be content with what we have and fix the predicted problems to make our planet earth a better place to live in.

Our ancestors harnessed the natural resources of the earth and beyond for science and technology. They were able to advance their technology and perfected space travel. Perhaps, just like the Tower of Babel in the Bible, during the time of King Nimrod, they almost reached the abode of God. Imagine if earth and beyond will disintegrate and disappear, what chain reaction could take place when the magnetic field of the earth will be disturbed? Perhaps, earth and beyond will stay intact but may drastically change its landscape in the end/beginning of the cycle but earth will continue to support life.

These magnificent unexplained structures found all over the world that are supposed to be made by Extraterrestrials or Aliens with the use of highly advance technology that were not supposed to be existing yet during those days, could be the making of our ancestors who were able to advance their technology and travel into Space during their civilization. This could also be the reason why buildings and structures built during those days are made of huge and strong materials such as block of stones to withstand the catastrophic effects that humans are expecting about to experience.

Before humans journeyed into space, they build structures like the pyramids and stone formations that could withstand the holocaust and could also serve as guide if ever they survive the journey and decide to return someday. The earth survived the destruction and continue to support life until today. I think, those who were able travel into the universes landed on some distant planet and were able to adopt in the place. Perhaps, some landed on the moon and converted it into a magnetic shield to protect earth from debris in the universe that made the surface of the moon full of craters. Since then, they watch over us and visit earth from time to time as UFO's (Unidentified Flying Objects) to help in anyways they can while waiting for the right time to come home or invite us to where they are.

I think, these Hieroglyph in ancient texts, Carvings, Drawings on stones, mountains, caves and these Strange Statues, were records done by our ancestors before they traveled into the universe. Those who were not able to travel but survived the holocaust, made their own recordings of what they saw and what they know. The face of the earth was drastically changed during the end and beginning of the cycle that some living things who survived suffered the consequences from radioactive fallouts. These freak creatures we call abomination, half man half animal could be the result of radioactive contamination from the catastrophic event. I think, what humans should do now is to unite all governments into one humanity with combined religion and science under one God.

Humanity will be working together as one people to make sure that Space Exploration will not be Space Invasion. Universal Distancing will be observed to acknowledge acceptance and respect to other inhabitants of other planets if there is any. Then perhaps, damage to earth and life on earth and beyond might be lessen and the cycle will slow down almost to a halt. Like the fountain of youth, life expectancy will be much longer until one gets bored living in this world without end, "so be it".

Instead of spending billions of US dollars on Space Explorations, mankind can work on projects to protect and improved the ecological balance of the earth and beyond. Mankind will find no need for destructive technologies and make use of technology to improve the life conditions of humans and find ways to end calamities, famines, and diseases. In short, put an end to all the things that are destructive to earth and harmful to life and improve the protection to mother earth and father sun. Perhaps, this way science might find what they are looking for, how to reverse the aging process of living things.

If Matter is a substance made up of various types of particles that occupies space, has mass and weight and Space is an empty boundless extent in which objects and events occur. I think, Matter cannot exist without Space, even Space cannot exist without the Space. God created all that is in the Space, but did God create the Space?

THOUGHTS

By the late: Manuel T. Enerva
Diyaryo Magazine 1997
"WHY I DON'T BELIEVE
IN THE EXISTENCE OF
THE HUMAN SOUL":

The soul as we understood from childhood is the other self, in spiritual form. Our religion – be it the Catholic, Protestant, Methodist, Buddhist, Muslim, or Hindu – taught us, that when a person dies, the soul lives in another world. Christianity believes that the soul answers for the good or evil that the person has done in his life, and from such actions are judge by the keeper of the portal of Heaven. If he has sought forgiveness for his sins before death then, depending on the quality of his life. Those who lived a lesser life, temporarily go to Purgatory, and those who lived a good life, immediately enters Heaven. The Hindus and the Buddhist believe in a spirit world where the soul dwells for a while, a place of transition from one life to another. In that place, the soul waits for a new body, a new life where he will continue the journey, in a cycle of life and death and re-birth until he finally perfects his life and enters a state of union with God, or pure spiritual state called "NIRVANA" by the Buddhists.

When we were a child there was a point in our lives when we learned of the reality of death. I, for one, like other men was totally disturb and terrified of the thought of death, and asked my mother the question: "Why am I going to die?" "But there is Heaven", my mother answered, "It is a perfect place of peace and joy in the presence of God". What I heard was the most satisfying response, a perfect answer to my query, which renewed my vigor, and carried me through adolescence with peace of mind. From then on, I never bothered to question death anymore.

In college, a student is trained to think, to reason and to experiment, hence, college education makes a philosopher of the student, and cultivates in him a scientific mind. He learns that reality could be explained by reason (philosophy) and by experimentation, induction, and deduction (science). A scientific mind questions every assumption, every truth or doctrine. When in doubt, as the mind makes a new explanation (hypothesis) and experiments. Hence, in college, we learned to distinguish between fact and fiction.

The belief in God or the Supreme Being is one of our valued cultural legacies and hard to get rid of. God as a product of our belief, came from the past expressed in symbols and they are a part of our daily lives. To go against such symbols could disturb the balance of our existence, psychological and even biological. Imagine how life could have been miserable to me, if in childhood, my mother told me that "God is dead," as what the philosopher Friedrich Wilhelm Niethzsche said. I could have been an extremely disturb boy, an unhappy and fearful individual, devoid of peace of mind.

To the believer, therefore, religion has a function. It makes him rid of fear of death. Man, since time immemorial, has always dreaded death. This fear of death, however, is relieved with the belief in the soul. The flesh dies, but the soul survives. The earliest civilization that believed in the existence of the soul is Egypt. The soul of the dead traveled to this afterworld. There, the gods judged the spirit and weighed the soul's heart against a feather that symbolized the truth. The Greeks believed in Elysium, a land at the end of the earth where the righteous went after death. Like the Christian Heaven, it was a place of "perfect happiness". On the other hand, they believed in a place underground called Hades or Hell a place of misery for the soul where the flesh wad destroyed by fire.

Death is the termination of the function of the human cells. Science, however, at present is gradually discovering the secrets of cells. One recent breakthrough of science is the cloning of any adult sheep by Scotch scientists Keith Campbell and Dr. Ian Wilmut. Through cloning, the scientists were able to create an exact duplicate of an animal like identical twins. Scientists predict that cloning of human beings is possible in the next one to ten years. From the scientific viewpoint, it could be irrational to say that a human being has been genetically duplicated by understanding nature of cells, science gives hope to man to conquer aging process, by bringing the dead back to life through the technology of cryonics.

If science succeeds in creating such technology, then the belief in the existence of the soul will be in jeopardy. If science would bring back the dead to life, then from the viewpoint of religion, the soul, which we believe to have departed the body and, in the afterworld, would be summoned back to life. Science, therefore, would be able to command the human soul back to life. But what if the soul is already in Hell, and has been punished by God?

Finally, my position is that the concepts of God, the afterlife, and the human soul are cultural legacies. We have taken such beliefs primarily from the Christianity, and it is very difficult for us to disregard such beliefs because they are deeply ingrained in our culture. The belief in God or the soul is a cultural illusion imposed on us out of necessity of our existence. It serves as a cure all for the fear of death and to give meaning to our life. Otherwise, life would be meaningless. But I do not say that religion alone can provide the state of mind which calmly accepts death. A non-believer can also do so through meditation or a right state of mind. Every individual has this capacity although a non-believer would require a greater effort than that of a religious mind.

I also maintain that consciousness is the product of the mind and the senses. The brain is the seat of self, an entity that is purely material. It makes being conscious of the world around him. When the body dies, the brain loses consciousness due to cellular deterioration. No other entity services the body, as religion claims the soul to be.

The development in genetic engineering, specifically cloning and cryonics puts the concept of the soul in lesser category. If in the future, humans are cloned, and if it is assumed that the soul exist in the clones, then man can produce souls in the scientific laboratory through cryonics, then man would bring back soul defying the final judgement of the soul's destiny. But from the scientific viewpoint, life begins with the cell and from cellular development spring being. Man is entirely a material being.

REINCARNATION

Christians believe that humans die only once before judgement day and will be resurrected because of Jesus Christ.

> *HEBREWS: Chapter 9;27 Just as it is appointed that human beings die once, and after this the judgement.*

Reincarnation and Resurrection both deals with life after death but differ in the manner of coming back to life or the process of re-existence. Resurrection is the belief that after death, the dead will rise like Jesus of Nazareth the Messiah and Reincarnation is the belief that when a living thing dies, the soul lives and comes back in another body or form. Reincarnation derived from Latin that literally means "to take on the flesh again" (physical body).

The religions that support Reincarnation are mostly from Asia specially Hinduism, Jainism, Buddhism and Sikhism, all originated from India. The Catholic Church and Christianity does not support Reincarnation but believes in Resurrection. In the Christian Doctrine of the Holy Trinity, the Father, the Son and the Holy Spirit are three divine persons in one God. The Son in the form of Jesus Christ, died for mankind, was buried and the third day was Resurrected from the dead. Islam and Christianity have similar views on death and the afterlife.

In a survey conducted by the Global Research Society and the Institute for Social Research, Independent Polling System of Society (IPSOS). Fifty One Percent (51%) of the people in the world believe in some form of afterlife. Twenty Three Percent (23%) believe in an afterlife, but not specifically in heaven or hell. Nineteen Percent (19%) believe in heaven and hell, another Seven Percent (7%) in ultimate Reincarnation and Two Percent (2%) believe in heaven but not hell.

Reincarnation gives the believers the assurance that after death the soul will come back in another body or form and can correct the mistakes from the previous existence to ensure union with God in the everlasting life and that it makes death less fearful. I think, life is a cycle, Birth, Existence and Death, and Reincarnation or Resurrection could be a part of it.

According to science, all living things have DNA or deoxyribonucleic acid. It is the hereditary self-replicating substance present in all living organisms that makes some of the similarities. This also means that with proper care and correct nutrient intake, all external and internal physical body parts of all living things are self-healing.

Human's DNAs carries the Genetic Code (Genes) that is responsible for making everyone uniquely different from one another. When a living organism dies, these Genes can survive for millions of years and when the dead returns to ashes, to dust, these Genes goes with it. I think, Reincarnation takes place when these Genes from the dead combines with the Genes of those about to be born. Changes maybe minor or significant, it could happen soon, late or never.

Some people experience strange things like seeing a person, a place, a thing or even an event for the first time yet so familiar. Or could feel, see or know things before it happens. Or could be in a situation as if it is a repetition of a previous occurrence that happened in the past with everything the same even the sequence of events, movements, scenery, colors and even the people around.

Perhaps, the effects of the combination of Genes will vary on the quality, compatibility and number of Genes that are combining. Perfect match of Genes to recreate an exact duplicate as in Resurrection of the deceased person is impossible yet possible. When the combination of these Genes is not good or not healthy, it produces what is called inborn abnormalities. Some of these abnormalities are good like persons with Photographic Memory, Mathematicians, Musicians, Geniuses and others. But others are bad like persons with Incomplete External or even internal Body Parts, Down Syndromes, Midgets, Giants and others. What is worst is when it produces freak abnormalities like a person, part human and part whatever, abomination.

Evolution is the process of gradual formation or growth or development that takes place during existence. Perhaps Charles Darwin's Theory of Evolution that humans evolve from ape could not be proven because the missing link could not be found. Maybe it could not be found and studied because it was destroyed or buried so deep during the end/start of an Earth Cycle. Or maybe the changes in the DNAs takes place during conception in the womb. Or maybe humans did not really evolve from ape. Alienists say that human DNAs were manipulated by some High Technology from Alien Beings. I think, evolution is true to some extent like color, shape, size and some anomalies but does not change the whole specie. Human is Human, Ape is Ape, and Apple is Apple.

SPACEISM

With knowledge from the existing estimated Four Thousand Two Hundred (4,200) religions and advance technology from research and experiments in science today, mankind still commits the same crimes and sins over and over again.

Religion is trying hard to keep a person alive through the soul and science is working hard to save mankind by reversing the process of aging that everyone aspires to find the fountain of youth.

I think it is not fear of God and highly advance technology that will solve this problem but for mankind to be more humane and understanding rather than wealthy and powerful.

"THERE IS NO RELIGION HIGHER THAN TRUTH", a slogan of the Theosophical Society in the Philippines that lead me to my Theory of Spaceism.

Genesis is the first Book in the Bible that traditionally credited to Moses as the author. But according to modern scholars, it was written hundreds of years after Moses supposed to have lived in the 6th and 5th centuries BC.

GENESIS: Chapter 1;1 In the beginning, when God created the heavens and the earth, 1;2 the earth was a formless wasteland, and the darkness covered the abyss, while a mighty wind swept over the waters". 1;3 Then God said, "Let there be light," and there was light. 1;4 God saw how good the light was. God then separated the light from the darkness. 1;5. God called the light "day," and the darkness he called "night." Thus, evening came, and morning followed the first day.

According to science, the universe is approximately Thirteen Point Eight (13.8) Billion Years Old, and the earth is about Four Point Fifty-Four (4.54) Billion Years Old. But according to the Bible, God created the heavens and the formless wasteland earth about Six Thousand (6,000) and Ten Thousand (10,000) Years Ago. This produced heated disagreements between religion and science because many scientists ignored the recorded history of the Bible. On the other hand, Religion claims that science believed in an inflated age of the universe.

I think, both religion and science are correct in their claims. When God created the heavens Six Thousand (6,000) and Ten Thousand (10,000) year ago, the earth in the universe was already existing hanging in space. What the author of the Book of Genesis could have meant by heavens is the atmospheric boundaries of the Planet Earth with its Sun and the Moon. The word abyss in the creation could mean the SPACE where the

"Primordial Atom" of the "Big Bang Theory" exploded creating the Universe. But like this "Primordial Atom," where was God before and during creation or the "Big Bang"?

God created everything but space is not a thing. Space is an empty place where everything can exist, and nothing can exist without a Space. If God did not create the Space and God was in the Space before and during creation. Who created the Space? If God created the Space, this would raise an unending question as to who created the Space where God was when he created the Space. Otherwise, God must have just pop out from nowhere in SPACE.

No one really knows how the universe come into what it is today. Perhaps, it must have just appeared, pop-out or sprang out in space because of the invisible Electro Magnetic force that exist but cannot be seen by the naked eye in Space.

If God has no beginning and has no end as religion claims. God must be somewhere, anywhere, everywhere in Space before and during creation. Otherwise, God could just be a concept of the human mind that exist only by Faith. Or maybe God and Space coexisted and could even be one and the same. If Religion is a system of faith and worship to an unseen (God), and Truth is an accepted scientific fact like the (Space), where God is Religion and Space is Truth. Then, "THERE IS NO GOD HIGHER THAN SPACE". This is my Theory of Spaceism.

DESIDERATA

GO PLACIDLY AMID THE NOISE AND HASTE, & REMEMBER WHAT PEACE THERE MAY BE IN SILENCE, as far as possible without surrender be on good terms with all persons.

Speak your truth quietly & clearly; and listen to others, even the dull & ignorant; they too have their story.

Avoid loud & aggressive persons; they are vexation to the spirit. If you compare yourself with others, you may become vain & bitter; for always there will be greater and lesser persons than yourself.

Enjoy your achievements as well as your plans. Keep interested in your own career however humble; it is a real possession in the changing fortunes of time.

Exercise caution in your business affairs; for the world is full of trickery. But let this not blind you to what virtue there is; many persons strive for high ideals; and everywhere life is full of heroism.

Be yourself. Especially, do not feign affection. Neither be cynical about love; for in the face of all aridity and disenchantment, it is perennial as the grass.

Take kindly the counsel of the years, gracefully surrendering the things of youth.

Nurture strength of spirit to shield you in sudden misfortune. But do not distress yourself with imaginings. Many fears are born of fatigue & loneliness.

Beyond a wholesome discipline, be gentle with yourself. You are a child of the universe, no less than the tress & the stars; you have a right to be here.

And whether or not it is clear to you, no doubt the universe is unfolding as it should. Therefore, be at peace with God, whatever you conceive Him to be; and whatever your labors & aspirations in the noisy confusion of life, keep peace with your soul. With all its sham, drudgery and broken dreams, it is still a beautiful world. Be cheerful. STRIVE TO BE HAPPY

CONCLUSION

Since time in memorial, humans have been trying to put an end to Sins and Crimes or at least minimize the occurrences, especially Mortal Sins and Heinous Crimes. Aside from God's laws, "Ten Commandments", humans have also installed laws to prevent them from happening again and again. But instead of getting fewer, it is getting worse.

Sin is an act against God's laws that affects a person especially its Spiritual Body. This can be forgiven or removed by repentance and penance. Crime is an act against the laws humans have put in place that affects a person especially its Physical Body. If found guilty, it is documented and recorded with monetary fine or jail time or both. In extreme cases life imprisonment and the worst is death.

I think, combining religion and science humans will find wisdom to stop lying and unit under one God that will represents the whole of humanity. This way humans will have a God that did not create man according to its own image and likeness that cause divisiveness. When people are divided, they tend to be racist, and the skin color will be a big deal, Vanity and Greed will cause to dominate one another. But sins and crimes, life and death, and the color of what is under the skin are all the same.

There are two types of Sin, the Original Sin, and the Personal Sin. To the Christians, Original Sin is inherited by all humans by birth because Adam and Eve disobeyed God's order and as a punishment, they were driven out of the Garden of Eden as sinners doomed to die.

> *GENESIS: 3;16 To the woman he said: "I will intensify the pangs of your childbearing; in pain shall you bring forth children. Yet your urge shall be for your husband, and he shall be your master." 3;17 To the man he said: "because you listened to your wife and ate from the tree of which I had forbidden you to eat, "Cursed be the ground because of you! In toil shall you eat its yield all the days of your life. 3;18 Thorns and Thistles shall it bring forth to you, as you eat of the plants of the field. 3;19 By the sweat of your face shall you get bread to eat, Until you return to the ground, from which you were taken; For you are dirt, and to dirt you shall return."*

> *PSALM: 51;7 True, I was born guilty, a sinner, even as my mother conceived me*

> *ROMANS: 5;12 Therefore, just as though one person sin entered the world, and*

through sin, death, and thus death came
to all, inasmuch as all sinned- 5;21 so
that, as sin reigned in death, grace also
might reign through justification for
eternal life through Jesus Christ our Lord.

Original Sin can only be removed through baptism believing that Jesus Christ the only son of God, became human, to redeem the sins of mankind by dying on the cross. But how can a newborn baby especially while still in the womb would know about these things to believe?

Personal Sins are committed by humans by choice thru Freewill. There are two kinds of Personal Sin, the Venial Sin and the Mortal Sin. Venial means pardonable, in Catholic Theology, it is a lesser sin that does not result in an eternal damnation in hell and complete separation from God's Saving Grace. Mortal means fatal and subject to death, in Catholic Theology, it is a gravely sinful act that if not repented can lead to damnation and separation from God's Saving Grace.

Catechism of the Catholic Church
(CCC): 1857. For a sin to be mortal, three
conditions must together be met: "Mortal
sin is sinning whose object is grave matter
and which is also committed with full
knowledge and deliberate consent. (CCC):
1859. Mortal sin requires full knowledge
and complete consent. It presupposes
knowledge of the sinful character of the
act, of its opposition to God's law. It also

implies a consent sufficiently deliberate to be a personal choice. Feigned ignorance and hardness of heart do not diminish, but rather increase, the voluntary character of a sin.

If a person is committing Venial Sins repeatedly, this person will become a habitual sinner that can easily be led to committing Mortal Sins. Like Petty Crimes if done frequently could lead to bigger and more serious crimes. So, it is wise to stop committing Venial Sins and Petty Crimes totally and immediately because tolerance will only make them worst.

During our journey in this life on earth, we are confronted with various choices. We could either be materialistic or spiritualistic depending on our personal choice. There are times that our choice is contrary to what our conscience dictates but the result would be favorable. There are also times that no matter how hard we try to be good, everything goes wrong. These makes us think and feel that our life on earth is controlled by some force, as if predestined. I think, events from our previous choices with faith and luck make things to happen that lead to this seemingly predestined situation. But all of these could have been changed or avoided if we did choose to do so.

In the Bible, there are verses that says humans has the freewill to choose between good and bad and there are also verses that says humans are predestined. If our future on earth is predestined, we do not have to struggle in life. We will just wait for our destiny to come, while making the most out of our existence and Freewill will only be a spice of life.

Some Bible verses of Freewill:

> *JOSHUA: Chapter 24;15 If it does not, please you to serve the Lord, decide today whom you will serve, the gods your fathers served beyond the River or the gods of the Amorites in whose country you are dwelling. As for me and my household, we will serve the Lord.*

> *GENESIS: Chapter 2;16 The Lord God have man this order: "You are free to eat from any of the trees of the garden 2;17 except the tree of knowledge of good and bad. From that tree you shall not eat; the moment you eat from it you are surely doomed to die.*

> *GALATIANS: Chapter 5;13 For you were called for freedom, brothers. But do not use this freedom as an opportunity for the flesh; rather, serve one another through love.*

Some Bible verses of Predestine:

> *JEREMIAH Chapter 1;5 Before I formed you in the womb I knew you, before you were born, I dedicated you, a prophet to the nations I appointed you.*

PROVERBS: Chapter 16;4 The Lord has made everything for his own ends, even the wicked for the evil day.

JOHN: Chapter 15;6 Anyone who does not remain in me will be thrown out like a branch and wither; people will gather them and throw them into a fire and they will be burned.

I think even the life of Jesus the Messiah was predestined.

ISAIAH: Chapter 7;14. Therefore the Lord himself will give you this sign, the virgin shall be with child, and bear a son, and shall name him Immanuel.

MATTHEW: Chapter 1;21 She will bear a son and you are to name him Jesus, because he will save his people from their sins.

JOHN: Chapter 1;3 All things came to be through him, and without him nothing came to be 1;4 through him was life, and this life was light of people; :14 And the word became flesh and made his dwelling among us, and we saw his glory, the glory as of the Father's only Son, full of grace and truth.

Freewill is the freedom to choose given by God to mankind to shape their own future in the series of "Cause and Effect". God is good, just, merciful, loving and understanding that through God's "Grace", the law of nature will protect earth and beyond and maintain its ecological balance as humans ask for God's help, seek its own destiny, and knock on all opportunities.

> *MATTHEW: Chapter 7;7 Ask and it will be given to you; seek and you will find; knock and the door will be opened to you.*

But we must understand and accept that the results of these is not always good and favorable. It will entirely depend on the person's knowledge, wisdom, luck, and faith. What is important, is for humans to accept and respect whatever the results may be and remember the knowledge of wisdom proclaimed by Jesus Christ, the "Golden Rule".

> *MATTHEW: 7;12 Do to others whatever you would have them do to you. This is the law and the prophet.*

I think, the only bad on earth and beyond is bad itself with good, because there is no bad without good. Sins and Crimes are bad that cannot exist without good and good cannot exist without bad. God and humans have put in place laws for mankind to stop committing sins and crimes. But laws are applicable and effective only when they are already committed, and it seldom deters. There are countless studies done to find solutions on how to deter humans from committing sins and crimes, but still, it is getting worse. I think, looking into the beginning of Sins and Crimes, humans will find wisdom because the end is in the beginning.

I think, Adam and Eve were not the first man and woman on earth but the first husband and wife that God has wed. In the Bible, there are two stories of God's creation.

First Story of Creation:

> GENESIS: 1;20 Then God said, "Let the water teem with an abundance of living creatures, and on the earth let birds fly beneath the dome of the sky." And so it happened: 1;21 God created the great sea monster and all kinds of swimming creatures with which the water teems, and all kinds of winged birds. God saw how good it was, 1;22 and God blessed them, saying, "Be fertile, multiply, and fill the water of the seas; and let the birds multiply on the earth. 1;23 Evening came and morning followed-the fifth day 1;24 Then God Said, Let the earth bring forth all kinds of living creatures: cattle; creeping things, and wild animals of all kinds. 1;25 God made all kinds of wild animals, all kinds of cattle, and all kinds of creeping things of the earth. God saw how good it was. 1;26 Then God said "Let us make man in our image, after

our likeness. Let them have dominion over the fish of the sea, the birds of the air, and the cattle, and over all the wild animals and all creatures that crawl on the ground." 1;27 God created man in his image; in the divine image he created him male and female he created them. 1;28 God blessed, them saying: "Be fertile and multiply; fill the earth and subdue it. Have dominion over the fish of the sea, the birds of the air, and all the living things that move on the earth. 2;4 Such is the story of the heavens and the earth at their creation.

Second Story of Creation:

GENESIS: At that time when the Lord God made the earth and the heavens-2;5 while as yet there was no field shrub on earth and no grass of the field had sprouted, for the Lord God had sent no rain upon the earth and there was no man to till the soil, 2;6 but a stream was welling up out of the earth and was watering all the surface of the ground-2;7 the Lord God formed man out of the clay of the ground and blew into his nostrils the breath of life, and so man became a living being. 2;8 Then the Lord God planted a garden in Eden, in the east, and he placed there

> *the man whom he had formed. 2;9 Out*
> *of the ground the Lord God made various*
> *trees grow that were delightful to look at*
> *and good for food, with the tree of life in*
> *the middle of the garden and the tree of*
> *the knowledge of good and bad. 2;15 The*
> *Lord God then took the man and settled*
> *him in the garden of Eden, to cultivate*
> *and care for it. 2;18 The Lord God said:*
> *"It is not good for the man to be alone.*
> *I will make a suitable partner for him."*
> *2;19 So the Lord God formed out of the*
> *ground various wild animals and various*
> *birds of the air, and he brought them to*
> *the man to see what he would call them;*
> *whatever the man called each of them*
> *would be its name. 2;20 The man gave*
> *names to all the cattle, all the birds of the*
> *air, and all the wild animals; but none*
> *proved to be the suitable partner for the*
> *man.*

Clearly this is an anomaly in the Bible. In the First Creation, it says that man was created after the animals were created. In the Second Creation, it says that man was created before the animals were created.

When God created Adam, all his senses are in perfect good working condition. When God ordered Adam not to eat the fruit from the tree of knowledge, he handed down to Adam the "Freewill" to eat or not to eat the fruit from the tree of knowledge of good and bad. Since the Bible has not mentioned of a direct order from God to Eve not to eat the fruit from the tree of knowledge of good and bad, Adam must have handed it down to Eve.

> *GENESIS: 2;15 The Lord God then took the man and settled him in the garden of Eden, to cultivate and care for it. 2;16 The Lord God gave man this order: "You are free to eat from the any of the trees of the garden 2;17 except the tree of knowledge of good and bad. From that tree you shall not eat; the moment you eat from it you are surely doomed to die.*

The serpent, the most cunning animal that the Lord God has made, tricked Eve into eating the fruit from the tree of knowledge of good and bad.

> *GENESIS: 3;1 Now the serpent was the most cunning of all animals that the Lord God has made. The serpent asked the woman, "Did God really tell you not to eat from any of the trees in the garden? 3;2 The woman answered the serpent: "We may eat of the fruit of the trees in the garden; 3;3 it is only about the fruit of the tree in the middle of the garden that God said, "You shall not eat it or even touch it, lest you die.*

The serpent sensed that Eve was confused when she

answered. Eve was thinking of the tree of knowledge of good and bad but was referring to the tree of life that was in the middle of the garden and even went farther by saying that by just touching it you will die. The serpent took advantage of this and mixed the tree of life with the tree of knowledge of good and bad.

> *GENESIS: 3;4 But the serpent said to the woman: "You certainly will not die!*

Referring to the tree of life and mix it with the tree of knowledge of good and bad.

> *GENESIS: 3;5 No, God knows well that the moment you eat of it your eyes will be opened and you will be like gods who know what is good and what is bad.*

The trick worked, Eve took and ate the fruit of the tree of knowledge of good and bad and shared it with Adam. Adam also eats the fruit from the tree of knowledge of good and bad unaware disobeying God's order. Clearly the serpent *lied* by mixing true and false (trulse) to make it appear to be true. Remember when Cain killed his brother Abel, Cain *lied* to God by denying what he did. Truly *Lie* is the root of all evil.

From the time Adam and Eve eat the fruit from the tree of knowledge of good and bad, they acquired the Original Sin. Their consciousness and senses were opened that they felt shame and fear realizing they were naked. As a punishment, God drove Adam and Eve out of the garden of Eden as sinners doomed to die. But

God so loved his creation that he installed Conscience. A transceiver with exclusive frequency between God and mankind to protect the Soul of a departed person. From then on, God predestined persons, places, things, and events to guide each soul to restore its rightful place in the afterlife.

I think, when the needs and wants of a person's senses are suppressed, the tendency of its frequency is to increase. The Physical Body starts to grow while still in the womb of the mother where suppression of the senses starts because the baby depends entirely on the mother. When the baby comes out into this world for the first time, suppression is even greater because we do not understand what they are trying to tell us. We only know we did the right thing when the baby stops struggling and crying.

As the Physical Body of the baby grows and the mind starts to think and communicate, the Spiritual Body starts to grow. Then, self-control, laws and all kinds of suppression comes into the mind. Then humans start committing sins and crimes because of laws that suppresses the senses. I think, since mankind knows what is good and what is bad, equipped with freewill, senses must be free. Humans must learn to accept and follow what their conscience dictates. Let all the senses be free from suppressions and let freewill decide while conscience guide the human mind to make the right decision with wisdom from the combination of religion and science.

The best way to show the combining of religion and science is in the beginning of mankind.

GENESIS: 1;27 God created man in his image; in the divine image he created him male and female he created them.

Clearly God created perfect male and female, other than that would be a birth defect or abnormality, like Lesbian, Gay, Bisexual, Transgender, Queer (LGBTQ) and any internal, external physical deformities and mental health issues.

ROMANS: 1;24 Therefore God handed them over to impurity through the lust of their hearts for the mutual degradation of their bodies. 1;25 They exchanged the truth of God for a lie and revered and worshiped the creature rather than the creator, who is blessed forever. Amen.

Note: LIE is mentioned again.

Romans: 1;26 Therefore, God handed them over to degrading passions. Their females exchanged natural relations for unnatural. 1;27 and the males likewise gave up natural relations with females and burned with lust for one another. Males did shameful things with males and thus received in their own passions the due penalty for their perversity. 1;28 And since they did not see fit to acknowledge God. God handed them

*over to their undiscerning mind to do
what is improper.*

*DEUTERONOMY: 22;5 A woman shall
not wear an article proper to a man,
nor shall a man put on a woman's dress;
for anyone who does such things is an
abomination to the Lord, your God.*

*1 CORINTHIANS: 11;14 Does not nature
itself teach you that if a man wears his
hair long it is a disgrace to him, 11;15
whereas if a woman has long hair it is her
glory, because long hair has been given
(her) for a covering?*

*1 CORINTHIANS: 6;19 Do you not know
that your body is a temple of the holy
Spirit within you, whom you have from
God and that you are not your own? 6;20
For you have been purchased at a price.
Therefore glorify God in your body.*

After all these things that are written, said and done,
Jesus Christ is depicted to have long hair and worn
women's clothing during the time of his teachings, but
no one actually knows the true looks of Jesus. God
allowed these birth defects to happen in religion and
God also allowed surgery in science. In fact, God
performed the first surgical transplant when he took a
bone from Adam's ribs to create a woman. Then, there

is the Freewill for humans to choose and fix these birth issues.

One may change the body organ by surgery to match the mental state. Or maintain the body organ but change the mental state to correct the defect. A lesbian may choose to keep both mental and physical defects and have a gay for a partner who also kept both defects. This way they can become husband and wife, a male and a female.

Mankind must understand that these birth anomalies need physical and spiritual understanding, acceptance, and help. These defects are not natural because they are results of man's creation through sex. The choice will be through freewill by the individual concerned that God gave to mankind allowing them to decide what to do with their own body and soul. God did not create anything that is bad, God's creations are all very good.

> *GENESIS: 1;31 God looked at everything*
> *he had made, and he found it very good.*
> *Evening came, and morning followed-the*
> *six day.*

Bible scholars agree that there is not a direct mentioned regarding abortion in the Bible but something about a pregnant woman was mentioned.

> *EXDOUS: 21;22 When men have a fight*
> *and hurt a pregnant woman, so that she*
> *suffers a miscarriage, but no further*
> *injury, the guilty one shall be fined as*

*much as the woman's husband demands
of him, and he shall pay in the presence
of the judges. 21;23 But if injury ensues,
you shall give life for life, 21;24 eye for
eye, tooth for tooth, hand for hand, foot
for foot, 21;25 burn for burn, wound for
wound, stripe for stripe.*

All on earth and beyond that has to do with death started with Adam and Eve when they disobeyed God's order. Survivors from disasters, calamities, famine, and the likes, thank God for sparing their properties and lives. But what about the victims? Did they not pray for God's help and protection too? Is God choosy that favors some from others? God really works mysteriously in a person's mind. Did God really create man in his own image and likeness? Or maybe the other way around, man's mind created God according to man's own image and likeness.

John of Patmos is the author of the Book of Revelation (apocalypse), the last book in the Bible where he wrote a warning.

*REVELATION: 22;18 I warn everyone
who hears the prophetic words in this
book: if anyone adds to them, God will
add to him the plagues describe in this
book. 22;19 and if anyone takes away
from the words in this prophetic book,
God will take away his share in the tree
of life and in the holy city described in*

this book. 22;20 The one who gives this testimony says, "Yes, I am coming soon. Amen! Come, Lord Jesus! 22;21 The grace of the Lord Jesus be with all.

John of Patmos was a Christian exiled to the island of Patmos during the persecution of the Christians by the Romans. Patmos is a small island in Greece where he wrote the "Book of Revelation" while in exile. Traditionally, John of Patmos and John the Apostle son of Zebedee, author of the Gospel of John in the New Testament in the Bible is believed to be possibly one and the same person. However, other Christian writers such as Dionysius of Alexandria and Eusebius of Caesarea and many Biblical Scholars discounted this possibility.

I am not an Anti-Christ or an Atheist, but since the Bible has two parts, the Old Testament and the New Testament, maybe the "Book of Revelation" was a warning intended for the New Testament and not for the entire Bible. It would be better to believe and follow the teaching of Jesus Christ without fear but by understanding what he claims.

When mankind unites as one people under one God speaking in one tongue like the time of King Nimrod in the Bible, Lies, Sins, Crimes, Vanity, Greed, Divisiveness and other words that tends to create humanitarian problems will all be dead words. Then, mankind can survive all domestic and foreign invaders. Space Explorations can be reviewed to make sure that it is not invasion. Universal Distancing can be implemented as a sign of acceptance and respect to all other lifeforms on earth and beyond.

Physicist Stephen Hawking warned that Higgs Boson also known as God Particle discovered in 2012, has the potential to destroy the universe. Or perhaps, end the universal cycle for a new universe, a beginning of another universal cycle. My theory is that God is Space and Space is God. Truly Space has no beginning and has no end. Whereas God, as we know, was an inspiration to the minds of the people who wrote the Bible and could also end in the minds of the people. Mankind must be mindful and thankful to SPACE for providing mankind with a wonderful and mysterious mind that thinks, imagines and that manifests in faith. Angels that guide and protect mankind, Jesus Crist for the knowledge and wisdom, and God Particle somewhere, anywhere, everywhere in SPACE that made all visible and invisible possible in SPACE.

End